Literacy in Science and Technology: Learning Station Activities to Meet CCSS, Grades 6–8

English Language Arts Series

Authors: Schyrlet Cameron and Suzanne Myers
Editors: Mary Dieterich and Sarah M. Anderson
Proofreader: Margaret Brown

ISBN 978-1-62223-459-2

Printing No. CD-404210

Mark Twain Media, Inc., Publishers
Distributed by Carson-Dellosa Publishing LLC

Table of Contents

To the Teacher

The Common Core State Standards address standards for English Language Arts, which includes standards for Literacy in History/Social Studies, Science, and Technical Subjects. *Literacy in Science and Technology: Learning Station Activities to Meet CCSS, Grades 6–8* is the latest addition to the Mark Twain Media *English Language Arts (ELA)* series.

In the ELA series, students in grades six through eight explore reading, writing, and language in a learning-station environment. Learning stations engage students in individual or small-group activities. Learning stations are an instructional strategy that can be used to target specific skills.

Each book in the ELA series features five or six units of study. Each unit has a teacher page that identifies the goal, states the standards, lists materials and setup for the activities, and provides instructions to be presented to students. Also, there are questions for opening discussion and student reflection. (Note: It is important for the teacher to introduce, model, or review the concepts or skills with the students at the beginning of each unit.)

Literacy in Science and Technology: Learning Station Activities to Meet CCSS, Grades 6–8 contains five units of study. Each unit addresses the Science and Technology literacy section of the ELA Common Core State Standards. The units can be taught within the science and technical subjects content area or can be used as interdisciplinary units with English Language Arts.

Each unit consists of four to six learning stations. The activity at each station is designed to create interest, provide practice, and stimulate discussion. Some activities have accompanying handouts. Whenever applicable, technology skills are integrated into the activity.

Books in the ELA Series

- ***Reading: Literature Learning Stations, Grades 6–8***
  The units focus on alliteration, rhyme, plot and setting, tone and mood, and poetry.
- ***Reading: Informational Text Learning Stations, Grades 6–8***
  The units focus on citing evidence, bias, point of view, propaganda techniques, organizational text structures, and text features.
- ***Writing Learning Stations, Grades 6–8***
  The units focus on fact and opinion, characterization, making inferences, proofreading, and dialogue.
- ***Language Learning Stations, Grades 6–8***
  The units focus on punctuation, dictionary usage, figurative language, roots and affixes, and word meaning.
- ***Literacy in History and Social Studies: Learning Station Activities to Meet CCSS, Grades 6–8***
  The units focus on analyzing text, making inferences, and citing textual evidence using primary and secondary sources.
- ***Literacy in Science and Technology: Learning Station Activities to Meet CCSS, Grades 6-8***
  The units focus on following multi-step procedures, analyzing textual and visual information, and summarizing textual information.

Common Core State Standards Matrix

English Language Arts Standards: Literacy in Science & Technical Subjects										
Units of Study	**RST.6-8.1**	**RST.6-8.2**	**RST.6-8.3**	**RST.6-8.4**	**RST.6-8.5**	**RST.6-8.6**	**RST.6-8.7**	**RST.6-8.8**	**RST.6-8.9**	**RST.6-8.10**
Periodic Table of Elements		X		X			X			
Electricity	X		X		X		X			
Volcanoes				X			X		X	
Science Lab		X	X			X				
History of American Space Exploration		X					X		X	

Teacher Page

Unit: Periodic Table of Elements

Goal: Students will be able to use a Periodic Table to locate information about specific elements.

Common Core State Standards (CCSS):

CCSS.ELA-Literacy.RST.6-8.2: Determine the central ideas or conclusions of a text; provide an accurate summary of the text distinct from prior knowledge or opinions.
CCSS.ELA-Literacy.RST.6-8.4: Determine the meaning of symbols, key terms, and other domain-specific words and phrases as they are used in a specific scientific or technical context relevant to grades 6–8 texts and topics.
CCSS.ELA-Literacy.RST.6-8.7: Integrate quantitative or technical information expressed in words in a text with a version of that information expressed visually (e.g., in a flowchart, diagram, model, graph, or table).

Materials List/Setup

Station 1: Summarizing Text (Activity); Organization of the Periodic Table (Handout)
Station 2: Element Symbols (Activity); The Periodic Table of the Elements (Handout)
Station 3: Locating Information From a Periodic Table (Activity); The Periodic Table of the Elements (Handout)
Station 4: Graphing Textual Information (Activity)
Station 5: *Element Research (Activity)

Activity: one copy per student
Handout: one copy per each student in a group

*Integration of Technology Skills

Opening: Discussion Questions (Teacher-Directed)

1. What is an element?
2. What is the purpose of a Periodic Table? (Answer: It is a way of organizing and classifying elements according to their properties.)

Student Instructions for Learning Stations

At the learning stations, activities will focus on the elements in a Periodic Table. Discuss your answers with other team members after completing each activity.

Closure: Reflection

The following questions can be used to stimulate discussion or as a journaling activity.

1. Why did Dmitri Mendeleyev create the first Periodic Table?
2. What information can you find on a Periodic Table?
3. How is a Periodic Table organized?
4. Why do you think there is a difference in the amount of scientific information displayed from one Periodic Table to another?

Name: ______________________ Date: ______________________

Station One: Summarizing Text

Directions: Read the "Organization of the Periodic Table" handout. Determine the main idea of the selection. What details support the main idea? Complete the graphic organizer.

Main Idea

Detail

Detail

Detail

Directions: Use the information in your graphic organizer to help write a summary of the "Organization of the Periodic Table" reading selection.

Name: ______________________ Date: ______________________

Station Two: Element Symbols

Directions: Use "The Periodic Table of the Elements" handout to locate the symbol for each element in the box. Write the symbol in its proper place in the blank periodic table.

Aluminum	Arsenic	Calcium	Californium	Carbon	Cobalt
Copper	Helium	Hydrogen	Iodine	Iron	Lead
Magnesium	Mercury	Neon	Nitrogen	Nobelium	Oxygen
Platinum	Radon	Silicon	Silver	Tungsten	Uranium

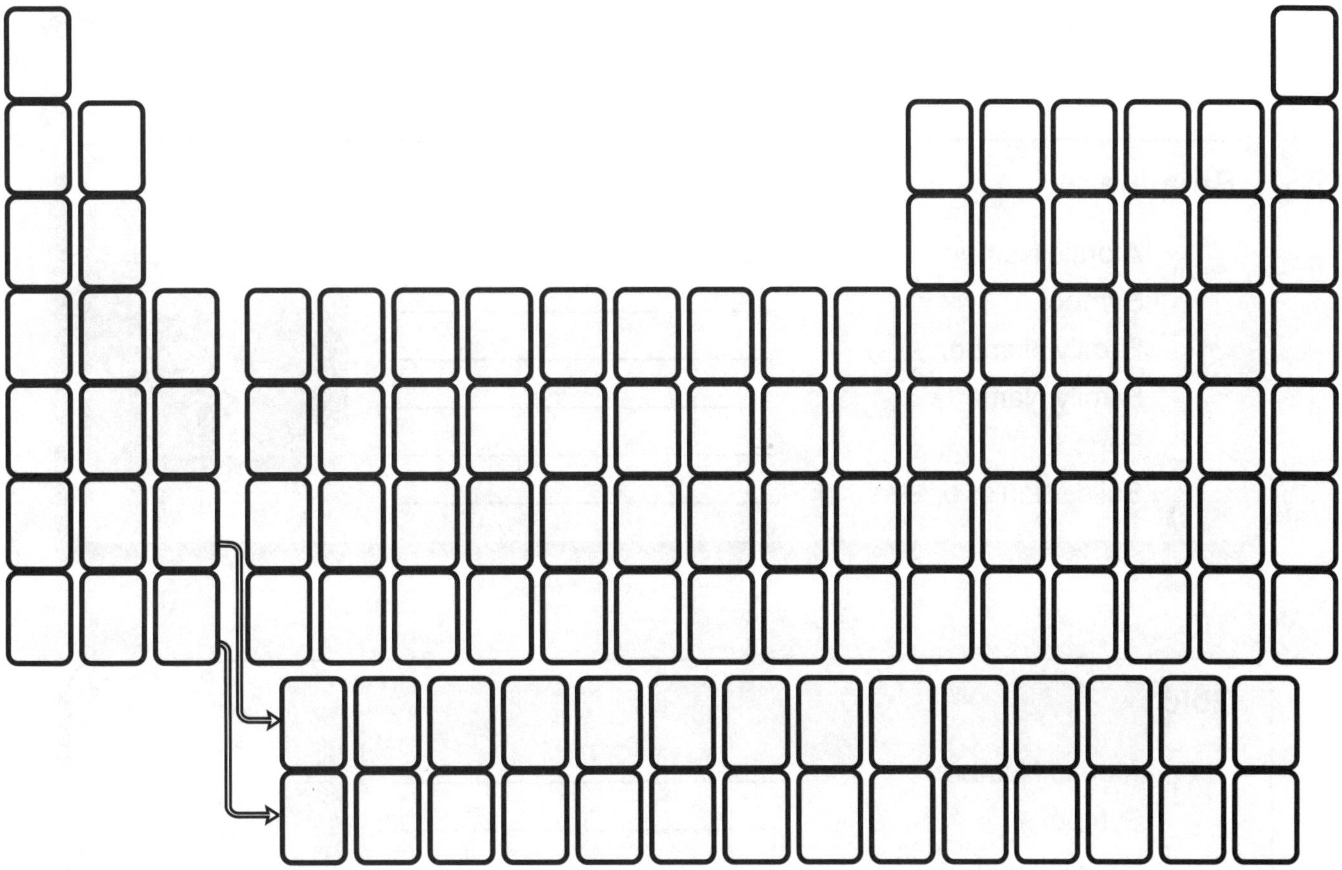

Name: ______________________ Date: ______________

Station Three: Locating Information From a Periodic Table

Directions: Use "The Periodic Table of the Elements" handout to locate scientific information about the elements oxygen, bromine, and gold.

Oxygen

Atomic Number ____________
Symbol ____________
Family Number ____________
Family Name ____________
Period ____________
Solid, Liquid, or Gas ____________

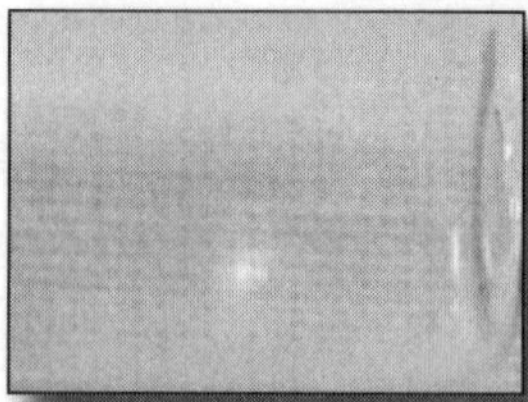

Bromine

Atomic Number ____________
Symbol ____________
Family Number ____________
Family Name ____________
Period ____________
Solid, Liquid, or Gas ____________

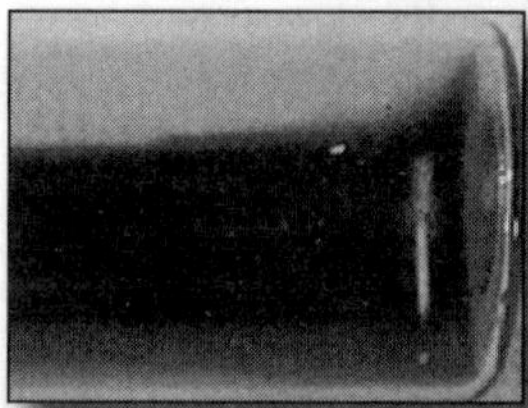

Gold

Atomic Number ____________
Symbol ____________
Family Number ____________
Family Name ____________
Period ____________
Solid, Liquid, or Gas ____________

Name: ______________________ Date: ______________________

Station Four: Graphing Textual Information

Directions: Read the paragraph and complete the pie chart.

Elemental Composition of the Human Body

The human body is primarily composed of four elements. Hydrogen, the most abundant element, makes up 63% of the total elements found in the body. Other abundant elements are oxygen (25.5%), carbon (9.5%), and nitrogen (1.4%). The remaining 0.6 percent is a combination of other elements.

Elements in the Human Body

Key

- ☐ Hydrogen
- ☐ Oxygen
- ☐ Carbon
- ☐ Nitrogen
- ☐ Other

Name: ______________________ Date: ______________

Station Five: Element Research

Directions: Select an element. On one face of the cube, copy the element's scientific information from a Periodic Table. Use print or online resources to find interesting facts about your element's history, appearance, properties, and uses. On the other faces of the cube, write the research facts. Cut out the pattern, fold, and glue tabs together to form a cube.

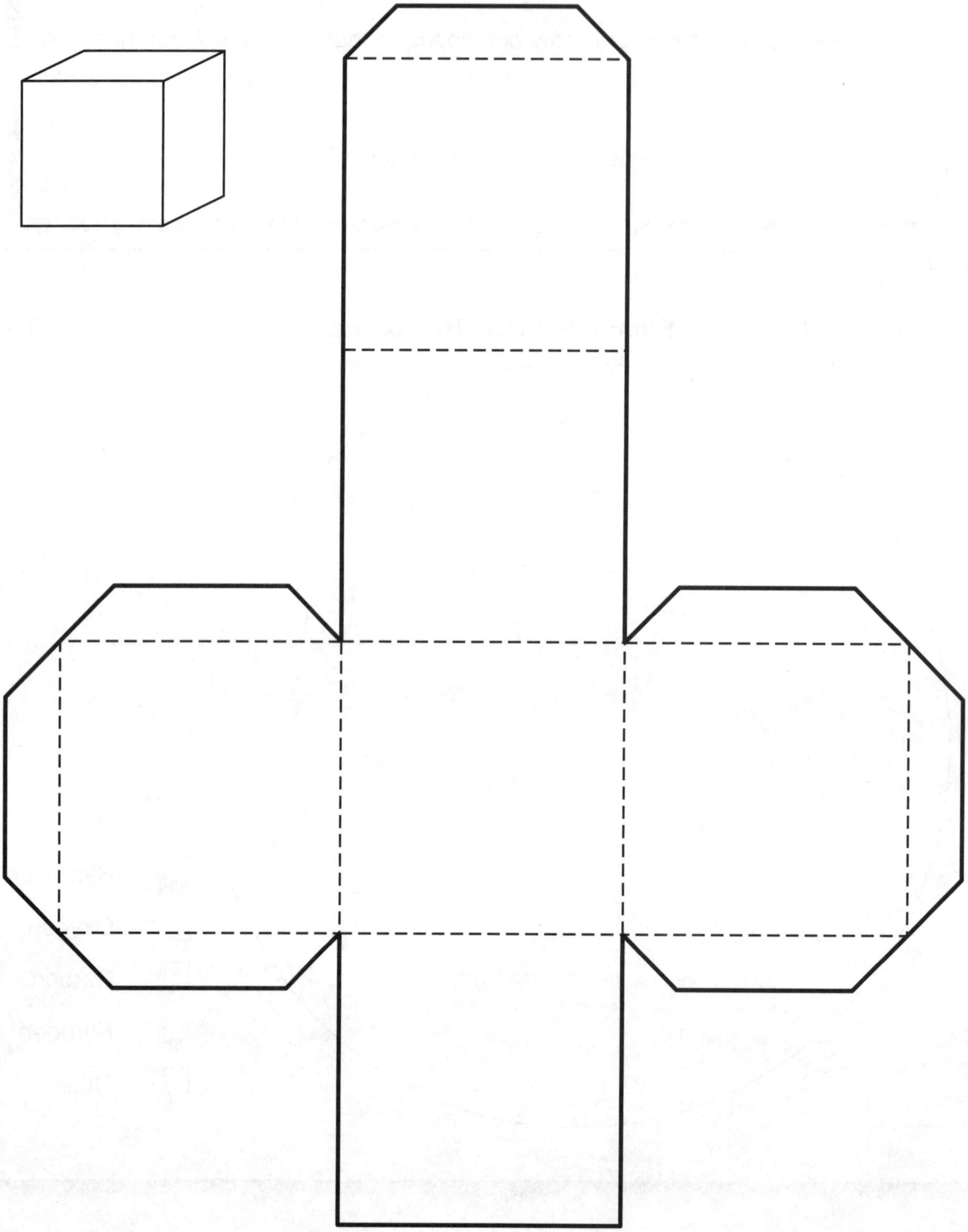

Handout

Organization of the Periodic Table

Dmitri Mendeleyev (1834–1907) was a Russian chemist credited with creating the Periodic Table in 1869.

In 1869, a Russian scientist by the name of Dmitri Mendeleyev constructed the first Periodic Table. Mendeleyev wanted to organize the elements according to their properties. His Periodic Table looked very much like the one used today. The Periodic Table used today not only organizes the elements by their properties, but it allows us to tell much about an element by its position on the table.

Organization

The Periodic Table is made up of horizontal rows called periods. The vertical columns of boxes are called families. Each box contains specific information about a single element. If you read the Periodic Table left to right and row by row, the atomic numbers will increase in order until you get to number 57. This is because elements 58–71 and 90–103 appear in order in the two rows at the bottom of the Periodic Table. They are called the lanthanides and actinides.

Key to Elements

Today, most Periodic Tables tell the element's name, symbol, atomic number, family, and other scientific information like atomic mass and electron configuration. Even though the amount and arrangement of information displayed on periodic tables may differ, they usually contain a key that tells us how to find the information we need. Many times you will see a key like this:

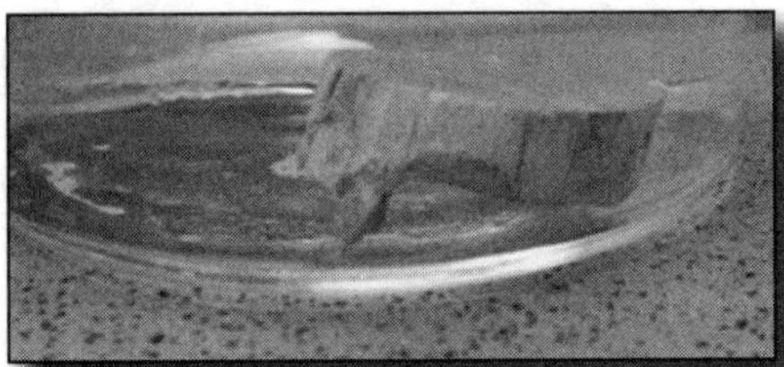

A sample of sodium

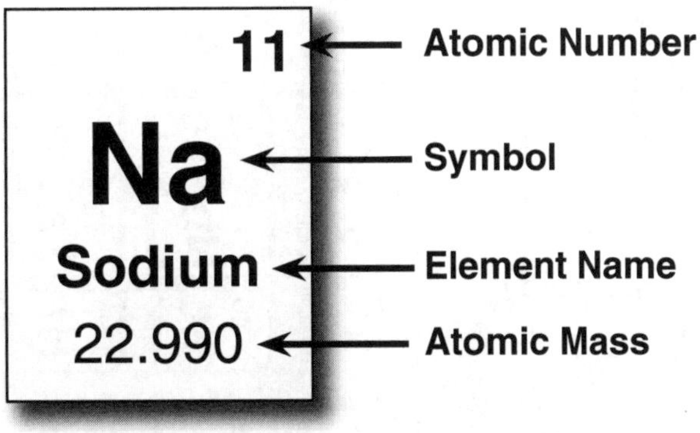

In the above example, the element's name is sodium. The symbol for sodium is Na. The symbol usually corresponds to the element's name. Symbols are usually two letters. However, some symbols have one or three letters. The atomic number for sodium is 11. This is the number of protons in its atom. A proton is a positively charged particle in an atom. The atomic mass is 22.990. This is the number of protons and neutrons added together.

Adapted from *Elements and the Periodic Table* by Theodore S. Abbgy. Used with permission of Mark Twain Media, Inc., Publishers

Handout

The Periodic Table of Elements

Key: 11 ← Atomic Number; Na ← Symbol; Sodium ← Element Name

Families

1	Alkali Metals
2	Alkaline Earth Metals
3–12	Transition Metals
13–16	BCNO Group
17	Halogens
18	Noble Gases

	1	2	3	4	5	6	7	8	9	10	11	12	13	14	15	16	17	18
1	1 H Hydrogen																	2 He Helium
2	3 Li Lithium	4 Be Beryllium											5 B Boron	6 C Carbon	7 N Nitrogen	8 O Oxygen	9 F Fluorine	10 Ne Neon
3	11 Na Sodium	12 Mg Magnesium											13 Al Aluminum	14 Si Silicon	15 P Phosphorus	16 S Sulfur	17 Cl Chlorine	18 Ar Argon
4	19 K Potassium	20 Ca Calcium	21 Sc Scandium	22 Ti Titanium	23 V Vanadium	24 Cr Chromium	25 Mn Manganese	26 Fe Iron	27 Co Cobalt	28 Ni Nickel	29 Cu Copper	30 Zn Zinc	31 Ga Gallium	32 Ge Germanium	33 As Arsenic	34 Se Selenium	35 Br Bromine	36 Kr Krypton
5	37 Rb Rubidium	38 Sr Strontium	39 Y Yttrium	40 Zr Zirconium	41 Nb Niobium	42 Mo Molybdenum	43 Tc Technetium	44 Ru Ruthenium	45 Rh Rhodium	46 Pd Palladium	47 Ag Silver	48 Cd Cadmium	49 In Indium	50 Sn Tin	51 Sb Antimony	52 Te Tellurium	53 I Iodine	54 Xe Xenon
6	55 Cs Cesium	56 Ba Barium	57 La Lanthanum	72 Hf Hafnium	73 Ta Tantalum	74 W Tungsten	75 Re Rhenium	76 Os Osmium	77 Ir Iridium	78 Pt Platinum	79 Au Gold	80 Hg Mercury	81 Tl Thallium	82 Pb Lead	83 Bi Bismuth	84 Po Polonium	85 At Astatine	86 Rn Radon
7	87 Fr Francium	88 Ra Radium	89 Ac Actinium	104 Rf Rutherfordium	105 Db Dubnium	106 Sg Seaborgium	107 Bh Bohrium	108 Hs Hassium	109 Mt Meitnerium	110 Ds Darmstadtium	111 Rg Roentgenium	112 Cn Copernicium	113 Uut Ununtium	114 Uuq Ununquadium	115 Uup Ununpentium	116 Uuh Ununhexium	117 Uus Ununseptium	118 Uuo Ununoctium

58 Ce Cerium	59 Pr Praseodymium	60 Nd Neodymium	61 Pm Promethium	62 Sm Samarium	63 Eu Europium	64 Gd Gadolinium	65 Tb Terbium	66 Dy Dysprosium	67 Ho Holmium	68 Er Erbium	69 Tm Thulium	70 Yb Ytterbium	71 Lu Lutetium
90 Th Thorium	91 Pa Protactinium	92 U Uranium	93 Np Neptunium	94 Pu Plutonium	95 Am Americium	96 Cm Curium	97 Bk Berkelium	98 Cf Californium	99 Es Einsteinium	100 Fm Fermium	101 Md Mendelevium	102 No Nobelium	103 Lr Lawrencium

Liquid - Br, Hg **Gas - H, He, N, O, F, Ne, Cl, Ar, Kr, Xe, Rn** **Solid - All Others**

Adapted from *Elements and the Periodic Table* by Theodore S. Abbgy. Used with permission of Mark Twain Media, Inc., Publishers

Teacher Page

Unit: Electricity

Goal: Students will be able to follow a multi-step procedure when conducting experiments or constructing models.

Common Core State Standards (CCSS):

ELA-Literacy.RST.6-8.1: Cite specific textual evidence to support analysis of science and technical texts.
ELA-Literacy.RST.6-8.3: Follow precisely a multi-step procedure when carrying out experiments, taking measurements, or performing technical tasks.
ELA-Literacy.RST.6-8.5: Analyze the structure an author uses to organize a text, including how the major sections contribute to the whole and to an understanding of the topic.
ELA-Literacy.RST.6-8.7: Integrate quantitative or technical information expressed in words in a text with a version of that information expressed visually (e.g., in a flowchart, diagram, model, graph, or table).

Materials List/Setup

Station 1: Text Features (Activity); Static Electricity (Handout)
Station 2: Following a Multi-Step Procedure (Activity); Electricity and Magnetism (Handout)
Station 3: Experimenting With Static Electricity (Activity)
Station 4: Transfer of Electrons (Activity); Static Electricity (Handout)
Station 5: Creating a Circuit (Activity); Electric Circuits (Handout)
Station 6: Constructing a Model (Activity); Electric Circuits (Handout)

Activity: one copy per student
Handout: one copy per each student in a group

Note to Teacher: All stations, except Station 1, will need materials and supplies. Quantities will vary depending on whether or not the activity is conducted individually or as a group.

Opening Activity and Discussion Questions (Teacher-Directed)

1. What is static electricity?
2. What is a conductor? Insulator?
3. What is an electrical circuit?

Student Instructions for Learning Stations

At the learning stations, you will explore the concept of electricity. Discuss the results with other team members after completing each activity.

Closure: Reflection

The following questions can be used to stimulate discussion or as a journaling activity.
1. Why are diagrams often included in multi-step procedures?
2. Draw a diagram of a series circuit with four bulbs. Create a key to the symbols used in the diagram.

Name: ______________________ Date: ______________

Station One: Text Features

Directions: Text features such as headings, captions, sidebars, and bullets are designed to help you quickly locate information. Read the "Static Electricity" handout. Use the information to answer each question. List the text feature that helped you locate the information.

Question	
1. What is the difference between conduction and induction?	**Answer:** **Text Feature:**
2. Do protons have a positive or negative charge?	**Answer:** **Text Feature:**
3. Is static electricity more likely to happen when the air is dry or humid?	**Answer:** **Text Feature:**
4. What is the relationship between lightning and static electricity?	**Answer:** **Text Feature:**

Name: ____________________ Date: ____________________

Station Two: Following a Multi-Step Procedure

Directions: Read the "Electricity and Magnetism" handout. The last paragraph of the handout describes a science experiment using a balloon and soda can. In the box below, list the materials and the step-by-step procedure you would follow to conduct this experiment. You may not need all of the lines provided. Conduct the experiment, record your results, and answer the question.

Balloon and Soda Can Experiment

Materials Needed: ____________________

____

Procedure:

1. ____________________
2. ____________________
3. ____________________
4. ____________________
5. ____________________

Results:

____

____

____

____

____

____

Did your results differ from the results stated in the handout? Explain.

____

____

____

____

____

Name: ______________________ Date: ______________

Station Three: Experimenting With Static Electricity

Directions: Conduct the Balloon and Cereal experiment. Record your observations in the data table. Write a conclusion for the experiment.

Balloon and Cereal Experiment

Materials Needed: O-shaped breakfast cereal
two pieces of string, 12 inches long
tape
two balloons
piece of wool material

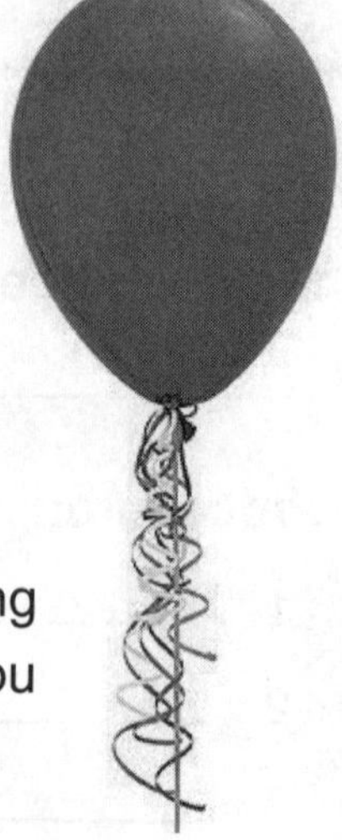

Procedure:

1. Gently tie the string to a piece of O-shaped cereal.
2. Tape the other end of the string to a surface that will allow the string to swing freely without coming into contact with another object. You might use the side of your table or a desk.
3. Rub a balloon against your hair.
4. Slowly, move the balloon close to the cereal that is dangling from the string.
5. Record your observations.
6. Repeat steps 1 and 2 with a new piece of cereal and string.
7. Rub a different balloon against a piece of wool cloth.
8. Slowly, move the balloon close to the cereal that is dangling from the string.
9. Record your observations.

Observations:

Balloon Against Hair	Balloon Against Wool

Conclusion:

____

____

____

____

Name: ____________________ Date: ____________________

Station Four: Transfer of Electrons

Directions: Read the "Static Electricity" handout. Conduct the following experiment. Record your results in the data table. Use the data and what you learned from the handout to write a conclusion.

Static Electricity Experiment

Materials Needed: five air-filled balloons, piece of wool, piece of paper, piece of silk, piece of plastic, stopwatch

Procedure:

1. Rub a balloon against your hair.
2. Immediately place the balloon against a wall and begin timing how long it stays stuck to the wall.
3. Record the time in the data table.
4. Rub a different balloon against one of the other objects in the data table. Repeat steps 2 and 3.
5. Repeat step 4 for all the items listed in the data table.

Results:

Object	Did it stick? (Yes or No)	How Long? (in seconds)
hair		
wool		
silk		
plastic		
paper		

Conclusion:

____

____

____

____

Name: ______________________ Date: ______________

Station Five: Creating a Circuit

Directions: Conduct the Lemon Experiment in Part A. Record your results. Read the "Electric Circuits" handout. Answer the questions in Part B.

Part A

The Lemon Experiment

Materials Needed: 6-cm piece of copper wire
small, non-coated, metal paper clip
lemon

Procedure:

1. Straighten out a paper clip. It should be the same length as the piece of copper wire.
2. Stick one end of the copper wire into the lemon.
3. Beside the copper wire, but not close enough to touch, stick one end of the paper clip into the lemon.
4. Touch the exposed end of the copper wire and the paper clip against your tongue at the same time.
5. Record your results.

Results:

____

____

____

____

Part B

1. What type of circuit did you create? ______________________

____

2. In this circuit, what is the battery? What is the switch? ______________________

____

3. This circuit does not create enough volts to power a penlight bulb. What could you add to the circuit to increase the voltage? ______________________

____

Name: ______________________ Date: ______________

Station Six: Constructing a Model

Directions: Follow the steps in the procedure to construct a model of a circuit. Then read the "Electric Circuits" handout and answer the questions.

Constructing a Circuit

Materials Needed: one 3″ x 5″ index card; 1.5-volt battery and battery holder; two brad fasteners; one paper clip; pieces of pre-stripped, insulated wire, 3–4″ each; penlight bulb and bulb holder

Procedure:

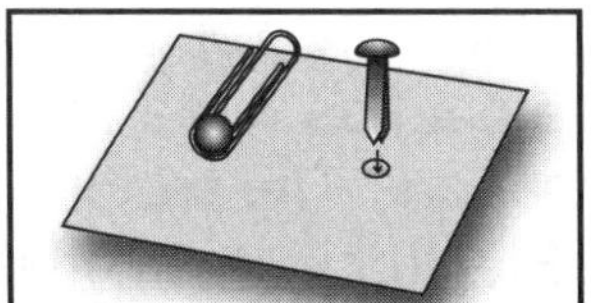

Figure A

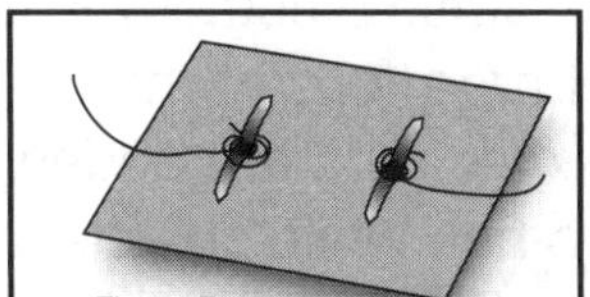

Figure B

1. Make a hole in an index card approximately one-third of the way from the end of one side of the card.
2. Make another hole about one-third of the way from the opposite side of the card.
3. Insert a brad fastener through one end of a paper clip and attach the brad to the index card. In the opposite hole, insert the second brad fastener. (See Figure A)
4. Rotate the paper clip to make contact with the second paper fastener. If it does not make contact, adjust the holes.

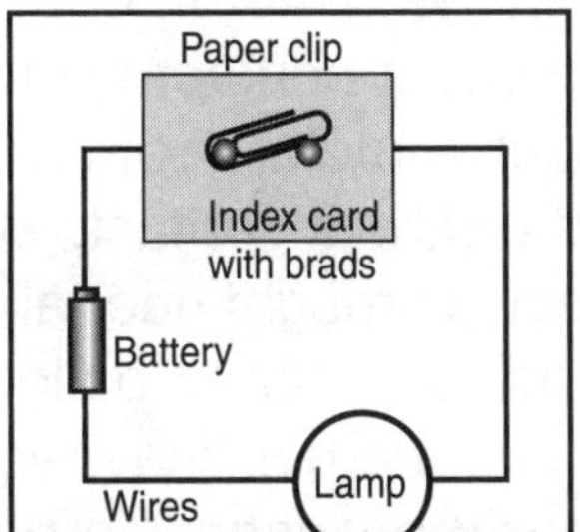

Figure C

5. Turn the card over on the backside. Attach a separate wire to the end of each brad fasteners. (See: Figure B)
6. Connect together the pre-stripped wires, battery, lamp, and the index card. (See Figure C)
7. To make the lamp light up, touch the paper clip against the opposite brad fastener.

1. What type of circuit did you create? Cite textual evidence to support your answer.

____

____

2. What is the function of the paper clip? Cite textual evidence to support your answer.

____

____

Handout

Static Electricity

Static electricity is electricity at rest. You are familiar with the effects of walking across a carpet and touching a doorknob or having a plastic bag stick to your clothes. These are examples of electric fields that are stationary or static, as opposed to a flowing charge or current electricity.

Electrostatic refers to electric charges that are confined to an object. This is also called static electricity or electricity at rest. It can also be defined as the charge-producing movement of electrons within a material and between materials. Electrons may move from atom to atom. Some materials hold electrons better than others. Plastic, cloth, and glass hold electrons tightly; they don't give them up. Other materials, such as metals, let electrons move more freely.

Basics of an Atom

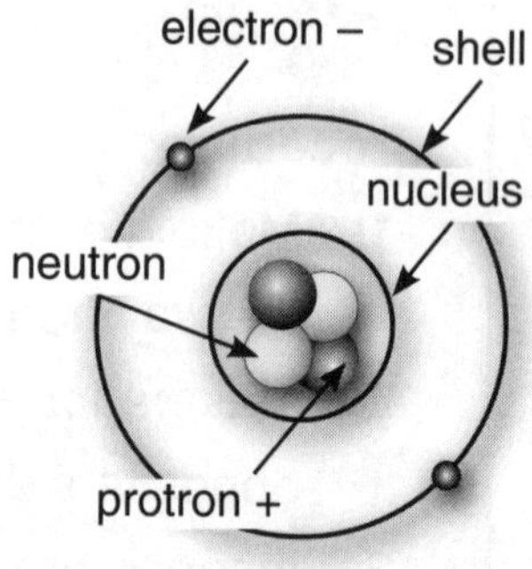

Understanding the basics of an atom, we can begin to see how static electricity works. All objects are made up of tiny parts, or atoms. Each atom has three tinier parts. These parts are protons, electrons, and neutrons. The protons and neutrons are located in the center of the atom. Electrons are in orbit around the nucleus.

How an Object Acquires a Static Charge

Static electricity is the result of the transfer of electrons. A positive charge forms when electrons are removed from a material, and a negative charge forms when electrons are added to a material. It can be observed that objects with like charges repel each other, while objects with opposite charges attract each other, and a charged object attracts a neutral object. Objects can acquire a static charge through friction, conduction, and induction.

- **Friction:** Objects can acquire charges (electrons) by rubbing one surface against another. We experience this sometimes when we walk across a carpet in the winter in our socks.
- **Conduction:** Charges (electrons) may be transferred through direct contact with other objects. This happens when you slide across a car seat and touch a metal door handle.
- **Induction:** A charge (electrons) may be induced in other objects without contact, with no transfer of electrons. An induced charge is when a charged object is brought near another material and the material responds by movement. When you rub an inflated balloon on a wool sweater and then bring the charged balloon near your hair, the hair may respond by standing up straight. In this case, no electrons are transferred between the balloon and the hair; instead, free electrons are being repelled, and the resulting positive charge toward the ends of the hair causes it to be attracted to the negatively charged balloon.

Dry Air and Static Electricity

Static electricity demonstrations and experiments work best in dry air. You may have observed that you are more prone to static discharge experiences in the winter, when the relative humidity in indoor spaces is lower. The reason for this is that charges naturally leak into the air, and humid air will cause the electrons to dissipate faster. Static electricity is more common in desert areas where the air is dry.

Did You Know?

Lightning is a discharge of static electricity.

Adapted from *Electricity and Magnetism* by Dr. John B. Beaver and Dr. Don Powers. Used with permission of Mark Twain Media, Inc., Publishers

Handout

Electricity and Magnetism

Masses attract each other with gravitational forces, but there are other attractive forces in the universe that are much stronger: the force between electrically charged particles is one. Such particles may also repel each other. The nature of these forces has to do with the way in which matter is put together with tiny particles called **atoms**.

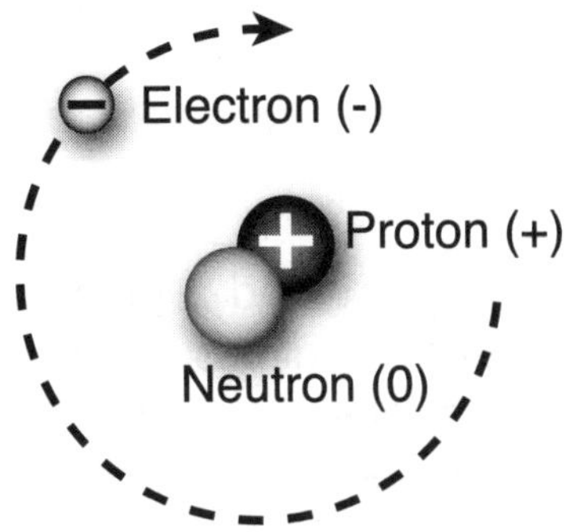

Three particles—protons, neutrons, and electrons—come together to create an atom. **Electrons**, which are very tiny, carry a negative charge. **Protons** are 1,800 times bigger than electrons, but they carry an equal, but opposite, positive charge. **Neutrons**, just slightly bigger than protons, are electrically neutral. Protons and neutrons contain most of the mass of an atom and form its center, or **nucleus**. Electrons orbit the nucleus in a kind of cloud of negative charge. Most matter contains an equal number of protons and electrons, making it electrically neutral, but electrons can be stripped from one atom and added to another. When this happens, the atom losing one or more electrons has a net positive charge, and the one gaining one or more electrons becomes negative. **Like charges** repel each other. **Unlike charges** attract.

If you rub a balloon on your hair, the balloon will remove some electrons. The balloon becomes negative, and your hair becomes positive. Hold the balloon a short distance from your hair (without touching it). The strands of hair reach for the balloon. The unit of charge is called a **coulomb** and represents about 6.28 billion electrons. While this number seems to be large, it is only about the amount of charge that passes through a 100-watt light bulb in one second.

Rub the balloon on your hair again, then stick it to a wall. The wall should be electrically neutral. Why the attraction? The negative charge of the extra electrons on the balloon is pushing away the electron clouds of the wall's atoms, separating the positive charge of their nuclei and the negative charge of the surrounding electron clouds. The balloon is attracted to the positive nuclei, which are now slightly closer to it. This process is called inducing an electric charge in a substance.

Rub that balloon on your hair again. Now place an empty aluminum soda can on its side on a smooth surface. Place the balloon near the side of the can at a low angle. You should be able to push the can along without touching it. You can induce an electrical charge in metals easily because they have loose outer electrons free to roam throughout the metal.

Adapted from *Science Tutor: Physical Science* by Gary Raham. Used with permission of Mark Twain Media, Inc., Publishers

Handout

Electric Circuits

Current electricity is the movement of electrons, which creates the flow of electricity. An **electrical circuit** or closed circuit is a complete path through which electrons flow from an energy source, through a conducting wire and appliance, and back to the energy source. **Switches** are used to open and close circuits. An open circuit is off, and a closed circuit is on. Electrical circuits are used to convert electrical energy into light, sound, and heat energy. There are different ways to connect appliances in a circuit.

Word Clue
The word **circuit** is related to the root word circle. A circle is represented by a closed figure; the circuit represents a continuous path through which electricity flows.

In its simplest form, a circuit must contain an energy source or battery and a piece of wire. Most often, the energy source is a dry-cell battery containing different chemicals that continuously react with each other, producing an excess of electrons. Hence, chemical energy is transformed into electrical energy. The chemicals commonly used in batteries are a mixture of ammonium chloride and zinc chloride, into which a carbon rod has been placed. The electrons leave the battery from the negative end of the battery (usually marked with a "-") and return to the positive end (usually marked with "+") through a wire called a conductor. Along the path of the wire, there may be one or more electrical appliances or devices, such as bulbs, motors, buzzers, etc.

Electrical circuits are either series or parallel. **Series circuits** require the electrical current to flow through all the devices in the circuit in just one path to make a complete circuit. **Parallel circuits** allow for more than one path along which the current can flow. Therefore, if one path in a parallel circuit has a break in it, the electrical current can still travel through the circuit that allows other devices in the circuit to operate. Because the current must flow through all the devices in a series circuit, the circuit has a higher resistance, and the bulbs in the circuit will be less bright. In a parallel circuit, because the current can flow around openings in the circuit, the resistance is less; therefore, the bulbs in the circuit are brighter.

Circuits are often shown with diagrams. To make diagrams consistent and easier to understand, electricians use symbols for certain objects. The illustration shows symbols that may be used to represent the electrical components in a circuit.

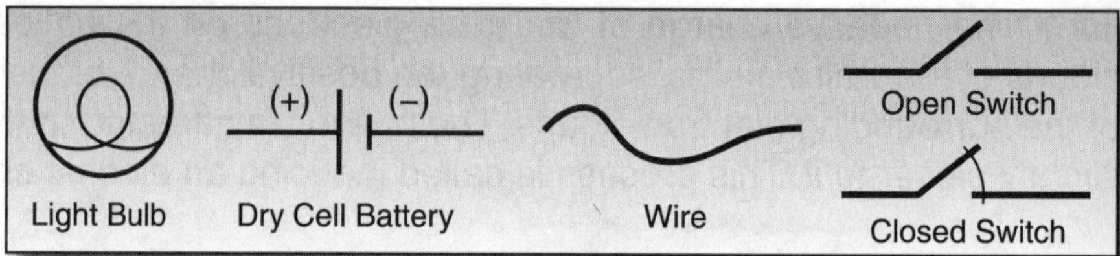

As electricity moves through a circuit, heat is generated. If there is too much demand on a circuit due to too many devices, excess heat will be generated. This heat could melt the insulation on wires and cause nearby substances to start a fire. Fuses and **circuit breakers** are devices in a circuit that will either melt or trip to prevent heat build-up.

Adapted from *Electricity and Magnetism* by Dr. John B. Beaver and Dr. Don Powers and *Confusing Science Terms* by Schyrlet Cameron and Carolyn Craig. Used with permission of Mark Twain Media, Inc., Publishers

Teacher Page

Unit: Volcanoes

Goal: Students will be able to analyze textual and visual information about volcanoes.

Common Core State Standards (CCSS):

ELA-Literacy.RST.6-8.4: Determine the meaning of symbols, key terms, and other domain-specific words and phrases as they are used in a specific scientific or technical context relevant to grades 6–8 texts and topics.
ELA-Literacy.RST.6-8.7: Integrate quantitative or technical information expressed in words in a text with a version of that information expressed visually (e.g., in a flowchart, diagram, model, graph, or table).
ELA-Literacy.RST.6-8.9: Compare and contrast the information gained from experiments, simulations, video, or multimedia sources with that gained from reading a text on the same topic.

Materials List/Setup

Station 1: Plate Tectonics: Analyzing Textual and Visual Information (Activity); Plate Tectonics (Handout)
Station 2: *Volcano Vocabulary (Activity); What Is a Volcano? (Handout)
Station 3: *Volcanoes in the United States (Activity); What Is a Volcano? (Handout)
Station 4: *Mount St. Helens (Activity)
Station 5: *Volcanic Past of the United States (Activity)

Activity: one copy per student
Handout: one copy per each student in a group

*Integration of Technology Skills

Opening Activity and Discussion Questions (Teacher-Directed)

1. What are some examples of visual information that can be found in a science textbook? (Answer: charts, diagrams, maps, data tables)
2. What is a volcano?

Student Instructions for Learning Stations

At the learning stations, you will integrate text and visual information to explore the subject of volcanology. Discuss the results with other team members after completing each activity.

Closure: Reflection

The following questions can be used to stimulate discussion or as a journaling activity.

1. Did combining textual and visual information help you have a better understanding of volcanoes? Explain your answer.
2. What did you learn about your state's volcanic past?

Name: ______________________ Date: ______________________

Station One: Plate Tectonics: Analyzing Textual and Visual Information

Directions: Read the "Plate Tectonics" handout. Use the information to answer the following questions.

1. What is the relationship between plate tectonics and volcanoes? Cite textual evidence.

____
____
____
____
____
____
____
____

2. What information about plate tectonics can you learn from the map? Explain your answer with examples or details.

____
____
____
____
____
____
____
____

Name: ______________________ Date: ______________________

Station Two: Volcano Vocabulary

Directions: Read the "What Is a Volcano?" handout. Go online to the Photo Glossary on the U.S. Geologic Survey Website. Study the photo for each vocabulary term listed in the graphic organizer below. Complete the chart using the textual information from the handout and the visual information from the Website.

Web Address: <http://volcanoes.usgs.gov/images/pglossary/index.php>
Title: USGS Photo Glossary of volcanic terms

Vocabulary Term	Definition	How did the photo help you better understand the vocabulary term?
Aa		
caldera		
cinder cone		
lava		
pahoehoe		

Name: ______________________ Date: ______________

Station Three: Volcanoes in the United States

Directions: Read the "What Is a Volcano?" handout. Go online to the National Atlas of the United States®. Read the directions on the opening Web page. Use information from the Website to complete the chart below.

Web Address: <http://www.nationalatlas.gov/dynamic/dyn_volcano.html>
Title: National Atlas of the United States®

Name	Location (State)	Type of Volcano	Dates of Latest Activity
1. Korovin	Alaska		
2. Yunaska	Alaska		
3. Akutan	Alaska		
4. Sunset Crater	Arizona		
5. Mount Shasta	California		
6. Clear Lake	California		
7. Long Valley	California		
8. Mauna Loa	Hawaii		
9. Haleakala	Hawaii		
10. Craters of the Moon	Idaho		
11. Mount Hood	Oregon		
12. Crater Lake	Oregon		
13. Mount Baker	Washington		
14. Mount Rainier	Washington		
15. Mt. Saint Helens	Washington		

Name: ______________________________ Date: ____________________

Station Four: Mt. Saint Helens

Directions: Go online to the U.S. Geological Survey Website listed below. Watch the video, *Mt. Saint Helens: A Catalyst for Change*. Then answer the questions.

Web Address: <http://gallery.usgs.gov/videos/235>
Title: *Mt. Saint Helens: A Catalyst for Change*

Question	Answer
1. What type of damage was caused by the eruption of Mt. Saint Helens in 1980?	
2. What are some new technologies that have been developed since May 18, 1980, for the use of volcano science and volcano monitoring?	
3. What are scientists studying using these new technologies?	
4. What areas are USGS observatories currently studying for volcanic activity?	

5. Reflection Question:
Summarize the effect the 1980 eruption of Mt. Saint Helens has had on the way scientists currently study and monitor volcanoes. Cite details and examples from the video to support your answer.

Name: ______________________________ Date: ____________________

Station Five: Volcanic Past of the United States

Directions: Go online to the U.S. Geological Survey Web page listed below. Under the heading "By States," click on your state. Locate six interesting facts about your state's volcanic past. Use the facts to complete the graphic organizer.

Web Address: <http://vulcan.wr.usgs.gov/LivingWith/VolcanicPast/>
Title: America's Volcanic Past, Volcanic Highlights and Features

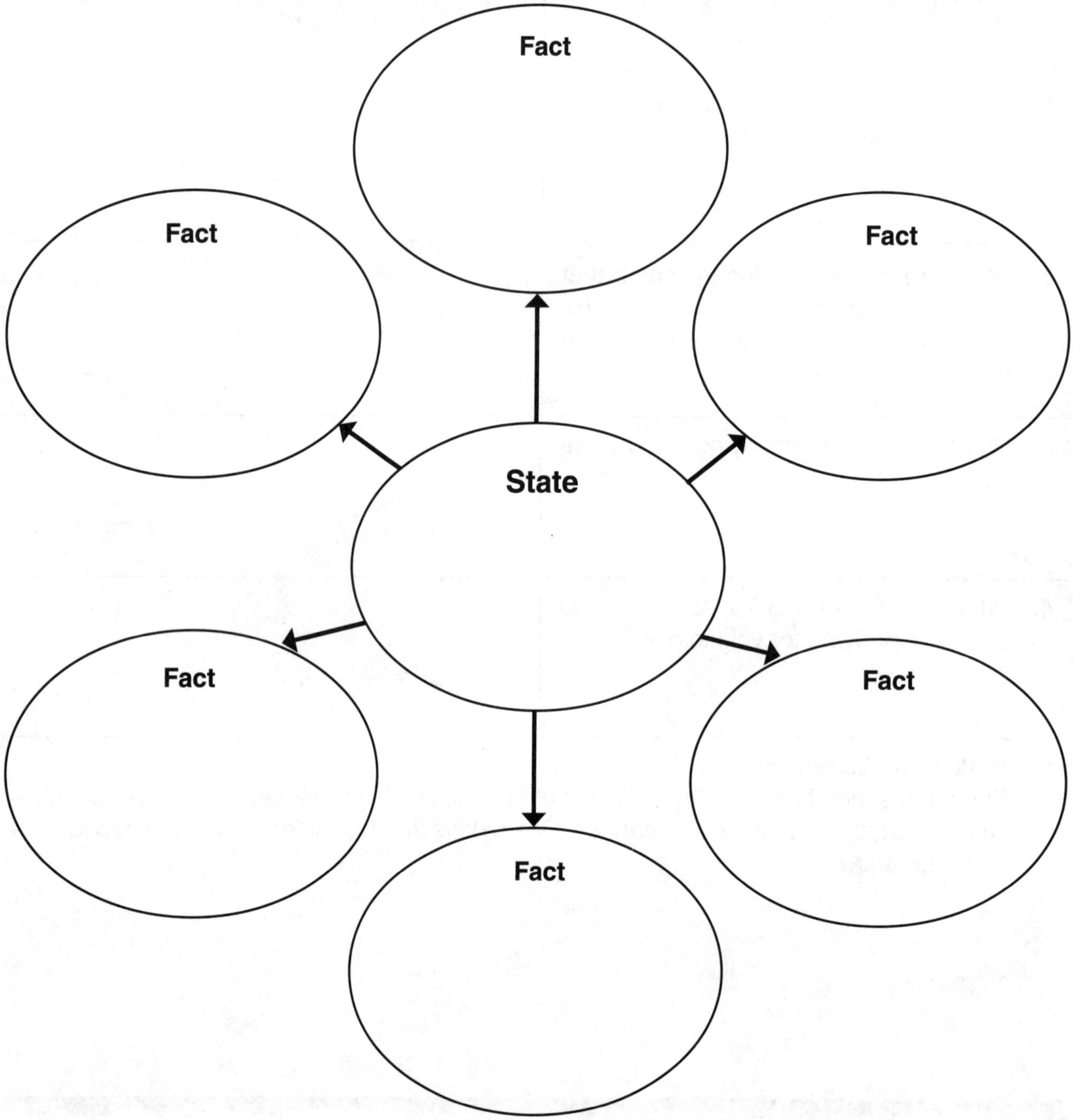

Handout

Plate Tectonics

Plate Tectonics and Continental Drift are both terms used to explain the present location of the continents. **Continental Drift** is the theory, proposed in the early 1900s, that states a supercontinent called **Pangaea** broke apart, forming the seven continents. The continents slowly drifted to their present positions. Today scientists use **Plate Tectonics** to explain how the continents were able to move to their present locations. Scientists believe the earth's outermost layer, the **lithosphere**, broke apart and formed seven large **plates** (pieces) and many smaller plates. The motion of magma in the mantle or **asthenosphere**, just under the crust, causes the movement of the plates.

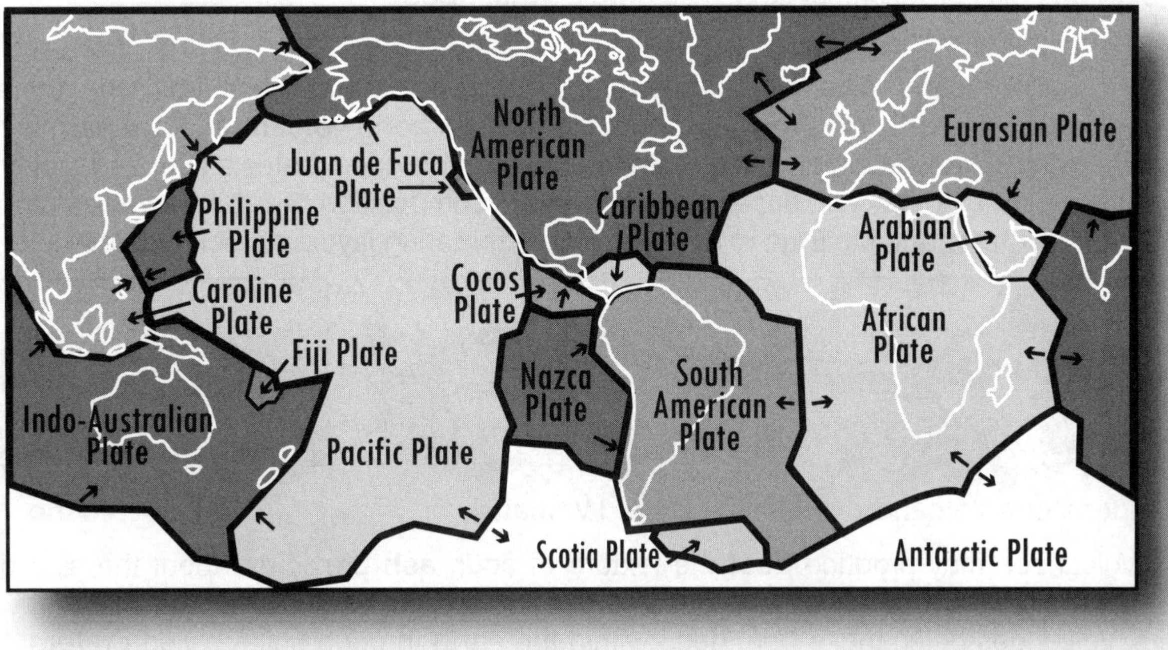

The plates move between 5 and 10 cm a year. The plates sometimes collide and grind past each other at the plate boundaries. Earthquakes, volcanic activity, mountain-building, and oceanic trench formation occur along plate boundaries. Three types of plate boundaries are convergent boundaries, divergent boundaries, and transform boundaries. **Convergent plate boundaries** crash together, often forming mountains like the Himalayas and producing earthquake and volcanic activity. **Divergent plate boundaries** move apart. Most divergent boundaries are under the ocean. They build undersea mountain ranges called mid-ocean ridges. **Transform boundaries** form where two plates slide past each other, often causing earthquakes.

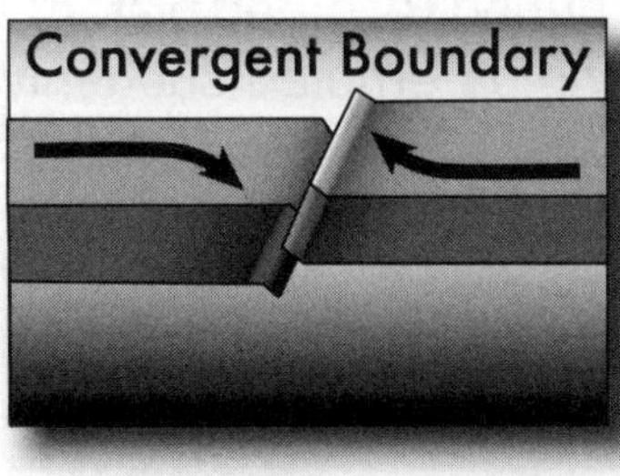

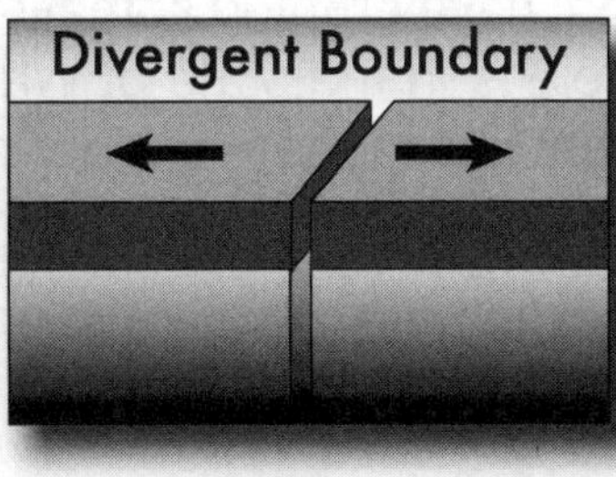

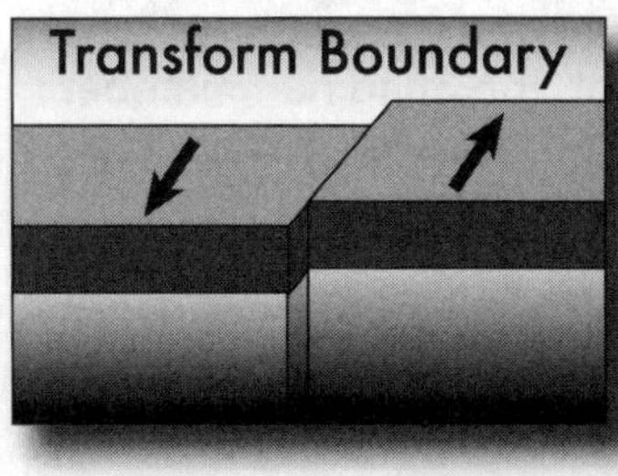

Adapted from *Geology: Rocks, Minerals, and the Earth* by LaVerne Logan. Used with permission of Mark Twain Media, Inc., Publishers

Handout

What Is a Volcano?

Volcanoes are exit holes at Earth's surface for molten rock called **magma** that finds its way to the surface by melting rocks above it or oozing through cracks. Once exposed at the surface, magma becomes **lava**. Hot, fast-moving dark lava with lots of iron and magnesium called **pahoehoe** (pah-HOH-ay-hoh-ay) solidifies into wavy, rope-like patterns. **Aa** (pronounced "Ah-ah,") is cool, slow-moving lava that forms spiky chunks. Both of these kinds of lava are produced during fairly slow and mild eruptions of dome-like **shield volcanoes** like Mauna Loa in Hawaii.

Aa flow from Mauna Loa

Violently erupting volcanoes often produce a light-colored lava high in silicon and aluminum compounds that becomes rhyolite, or granite. **Cinder cone volcanoes** have narrow bases and steep sides. Lava containing lots of water and carbon dioxide creates a light and "holey" rock called either **pumice** or **scoria**, depending on its exact composition. **Stratovolcanoes** (composite), like Mt. Vesuvius and Mt. Etna in Italy, produce alternating layers of rock and lava.

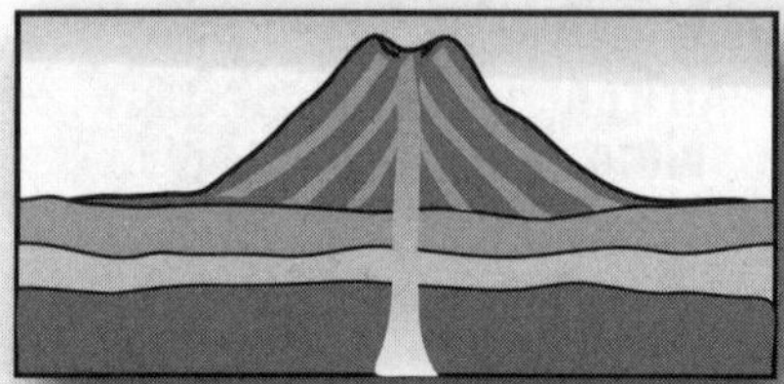

Cinder Cone Volcano

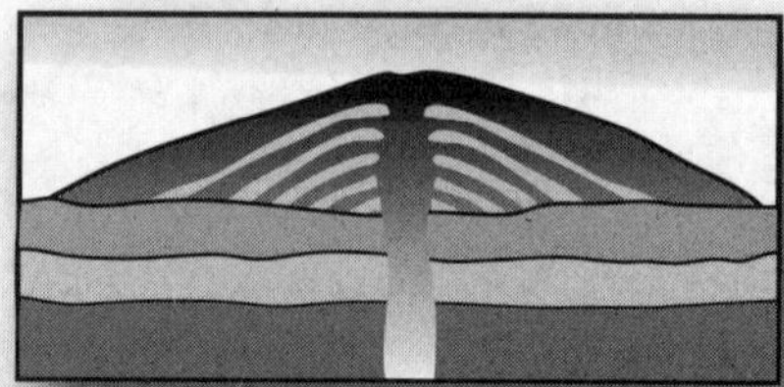

Shield Volcano

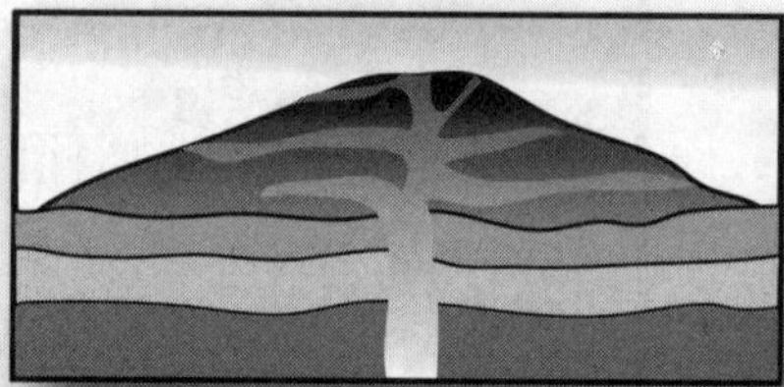

Stratovolcano

Volcanoes may produce **dust** the texture of flour, **ash particles** about the size of rice grains, or shoot out **volcanic bombs** the size of boulders. Golf-ball-sized projectiles are called cinders. Volcanoes belch these from deep within the earth through holes called **craters**. Lakes sometimes form in the basins of craters of inactive volcanoes. A broad crater three times wider than it is deep is referred to as a **caldera**.

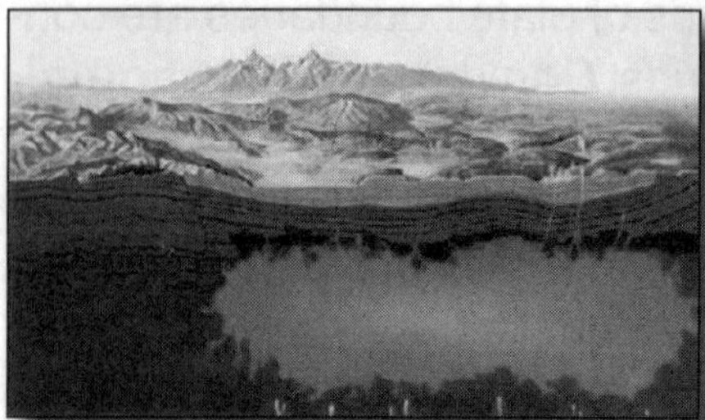

Drawing of the magma chamber believed to be under Yellowstone National Park

One of the most impressive calderas is the entire area of Yellowstone National Park in Wyoming. This ancient volcano is currently **dormant**, but it shows evidence of several past eruptions. It's currently past due for an eruption. **Extinct** volcanoes show no activity and are often weathered down to their old magma chambers. **Active** volcanoes are those that have erupted sometime during recorded history. Mt. Saint Helens in the state of Washington erupted suddenly in 1980 after many years of dormancy.

Slowly erupting volcanic areas in Earth's past have produced vast areas of layered lava extending over thousands of square miles. These areas are called **volcanic traps**. An example of this would be the Deccan Traps of southern India.

Adapted from *Science Tutor: Earth & Space Science* by Gary Raham. Used with permission of Mark Twain Media, Inc., Publishers

Teacher Page

Unit: Science Lab

Goal: Students will gain an understanding of science lab safety rules and procedures and how to use basic science equipment.

Common Core State Standards (CCSS):

ELA-Literacy.RST.6-8.2: Determine the central ideas or conclusions of a text; provide an accurate summary of the text distinct from prior knowledge or opinions.

ELA-Literacy.RST.6-8.3: Follow precisely a multi-step procedure when carrying out experiments, taking measurements, or performing technical tasks.

ELA-Literacy.RST.6-8.6: Analyze the author's purpose in providing an explanation, describing a procedure, or discussing an experiment in a text.

Materials List/Setup

Station 1: Lab Safety (Activity)
Station 2: Lab Equipment (Activity)
Station 3: Using a Compound Microscope (Activity); Magnification Power (Handout)
Station 4: Reading a Graduated Cylinder (Activity)
Station 5: Reading a Triple-Beam Balance (Activity)
Station 6: Analyzing Data (Activity)

Activity: one copy per student
Handout: one copy per each student in a group

Opening Activity and Discussion Questions (Teacher-Directed)

1. What is the name of a tool or piece of equipment used by scientists?
2. What type of science tools or equipment have you used before?
3. What are some safety procedures you should know before working in a science lab?

Student Instructions for Learning Stations

At the learning stations, you will learn about science lab safety rules and procedures and practice the correct usage of science lab equipment. Discuss your answers and/or lab results with other team members after completing each activity.

Closure: Reflection

The following questions can be used to stimulate discussion or as a journaling activity.
1. Why is it important to know safety procedures before doing lab activities?
2. Why do you think the author wrote the "Magnification Power" handout?

Name: ______________________ Date: ______________________

Station One: Lab Safety

Directions: Write the meaning and safety procedure for each symbol in the chart.

Safety Symbol	Meaning	Safety Procedure

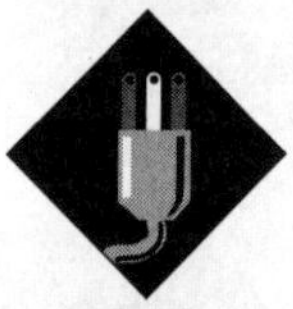		

Name: ______________________________ Date: ______________________

Station Two: Lab Equipment

Directions: Write the name of the lab equipment under the correct picture. Use your science book if you need help.

Word Bank

beaker	forceps	graduated cylinder	horseshoe magnet
magnifying glass	microscope	pipette	ring stand
safety goggles	spring scale	test tube	triple-beam balance

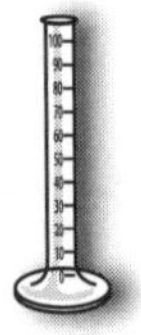 1. ____________	 2. ____________	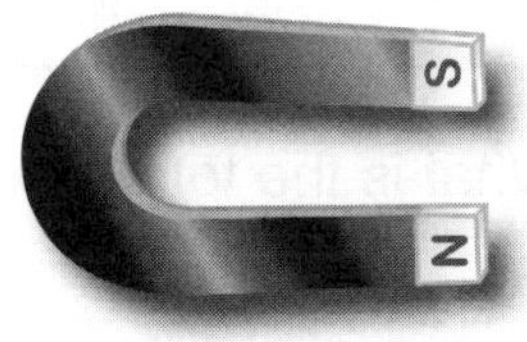 3. ____________
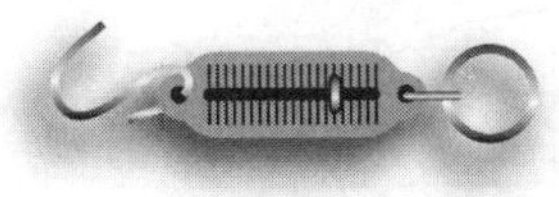 4. ____________	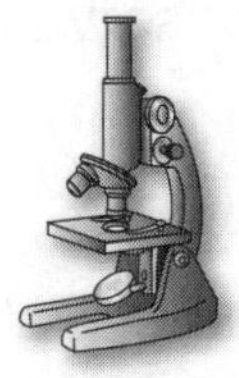 5. ____________	 6. ____________
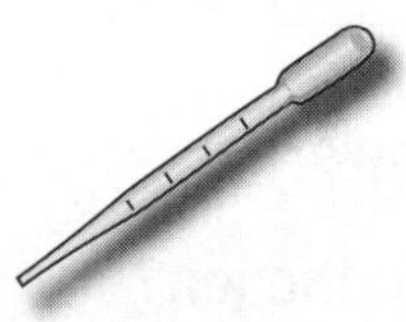 7. ____________	 8. ____________	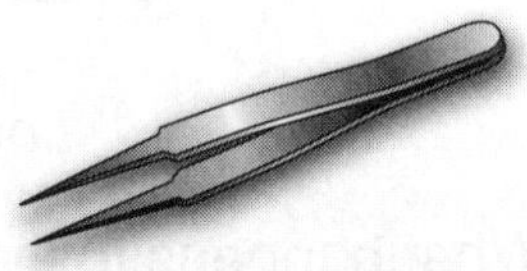 9. ____________
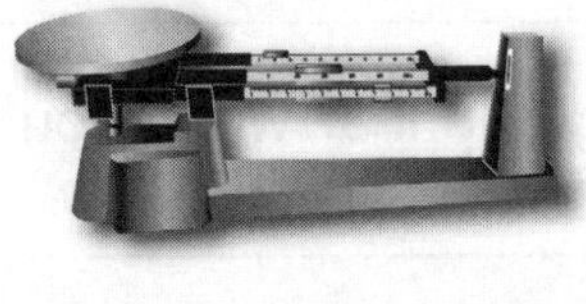 10. ____________	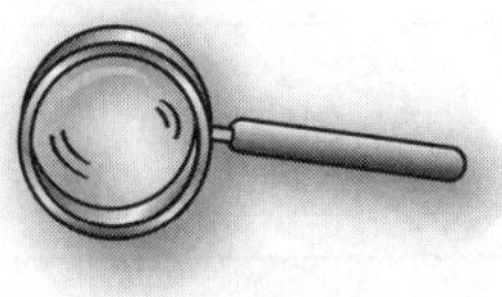 11. ____________	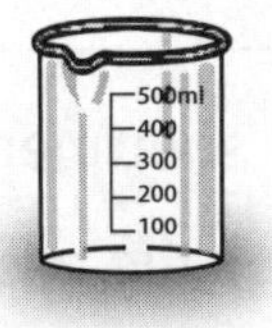 12. ____________

Name: ______________________ Date: ______________

Station Three: Using a Compound Microscope

Directions: Use a compound microscope and the "Magnification Power" handout to answer the following questions.

1. What is the power of the ocular lens? ______________
2. What is the power of the low-power objective lens? ______________
3. What is the power of the high-power objective lens? ______________
4. What is the total magnification power of the ocular lens and the low-power objective lens?

____

5. What is the total magnification power of the ocular lens and the high-power objective lens?

____

Directions: Write the first letter of your last name on a slide. Place the slide on the stage of the microscope with the letter right side up. Center it in the field and focus. View the slide on low power; then view on high power. Draw your observations.

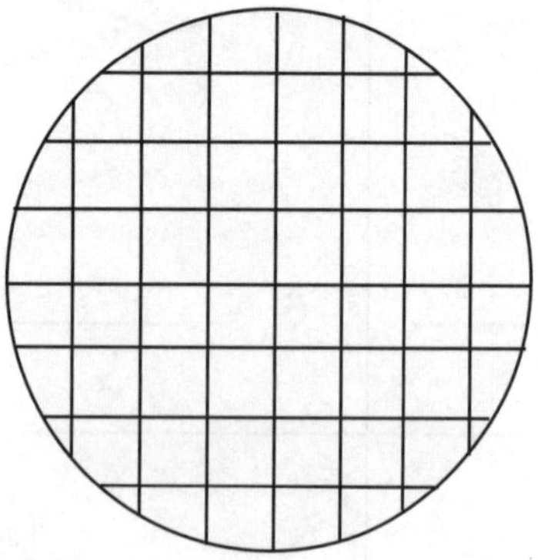

Low Power

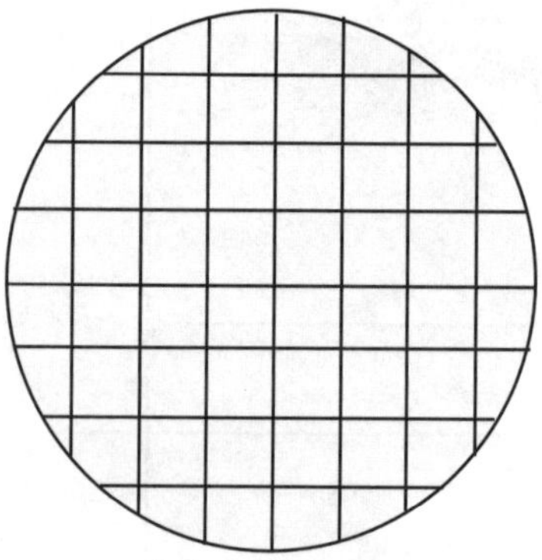

High Power

7. What happens to the image when you push the slide to the right? To the left?

____

8. What happens to the image when you push the slide up and away from you?

____

9. If the specimen on the slide is right side up, how will the image appear when you look through the microscope? ______________
10. Based on your observations, what can you conclude about viewing images with a compound microscope? ______________

____

Name: ______________________ Date: ______________

Station Four: Reading a Graduated Cylinder

Reading a Graduated Cylinder

Place the graduated cylinder on a flat surface such as a table or counter top. Your eyes should be directly level with the liquid. The liquid will tend to curve downward. This curve is called the **meniscus**. Read the volume at the bottom of the meniscus.

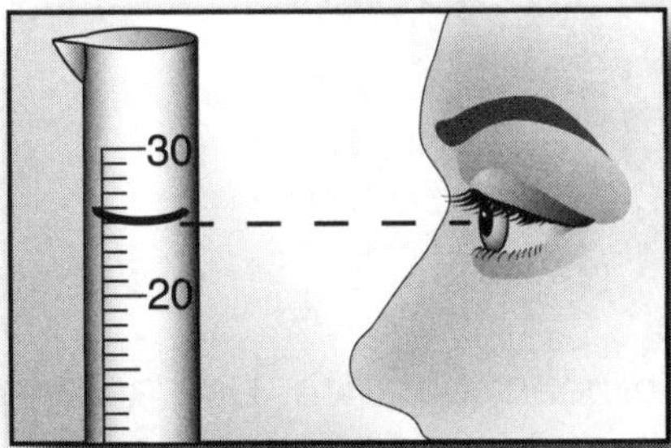

Measuring With a Graduated Cylinder

Liquid volume is measured using a graduated cylinder. The metric unit used is **milliliter** (mL). Graduated cylinders come in a variety of sizes including 10 mL, 25 mL, 50 mL, and 100 mL.

A graduated cylinder is **calibrated.** This means that the cylinder is divided or marked with lines forming a scale. The lines are the same distance apart from one mark to the next. There are a variety of scales used to label graduated cylinders. It is important to determine the scale (increments or value of each mark on the scale) before using it.

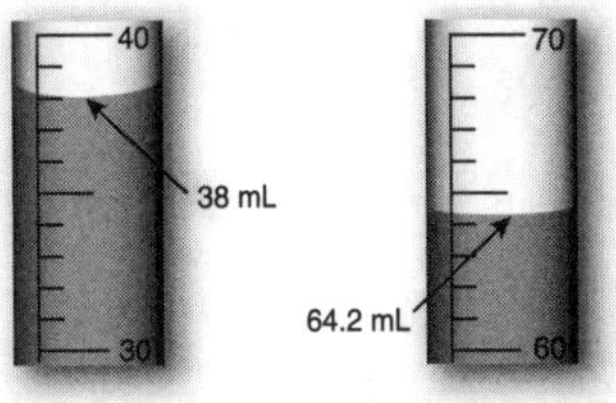

If the volume is between two marks on the scale, record the reading using a decimal.

Directions: Follow the steps below to find the volume of irregularly shaped objects. Record the water level and volume in milliliters (mL).

Step 1: Add water to a graduated cylinder. Record the water level in the Beginning Water Level column.
Step 2: Drop an object into the graduated cylinder.
Step 3: Record the new water level in the Ending Water Level column.
Step 4: Calculate the volume of the object. Subtract the Beginning Water Level measurement from the Ending Water Level measurement. Record your answer in the Volume column.
Step 5: Repeat Steps 1–4 for each object.

Object	Beginning Water Level (mL)	Ending Water Level (mL)	Volume (mL)
1.			
2.			
3.			
4.			
5.			

Adapted from *Scientific Method Investigations* by Schyrlet Cameron, Carolyn Craig, and Sherryl Soutee. Used with permission of Mark Twain Media, Inc., Publishers

Name: ______________________ Date: ______________

Station Five: Reading a Triple-Beam Balance

The triple-beam balance has one pan and three beams. One beam is the 500-gram (g) beam. It is divided into units of 100 g. Another beam is the 100-g beam. It is divided into units of 10 g. The front beam is the 10-g beam. It is divided into grams with each gram divided into tenths of a gram.

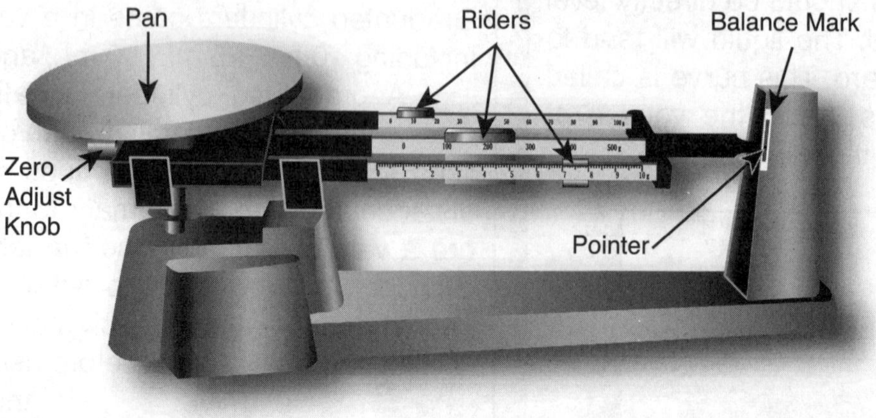

Directions: Follow the steps below to practice using a triple-beam balance. Use a variety of objects.

Step 1: Place the balance on a level surface and check to make sure the pan can move freely. Position all the riders at zero. Use the adjustment knob if the pointer does not come to rest in the middle of the scale.
Step 2: Place an object on the pan.
Step 3: Move the rider along the 500-g beam to the right until it is at the last notch that does not move the pointer below the zero point in the center of the scale. If you reach the last notch on this beam and the pointer is still below the zero point in the center of the scale, go to the next rider.
Step 4: Repeat Step 3 with the next rider and then, if needed, with the 10-g rider.
Step 5: Total the readings on all the beams to determine the mass of the object. Record the mass in the data table in grams.

Object	Mass (g)
1.	
2.	
3.	
4.	
5.	

Name: ____________________ Date: ____________

Station Six: Analyzing Data

In science, **work** is defined as exerting force on an object that causes the object to move in the same direction in which the force is exerted. You do work when you lift a box of books from the floor and place it on a table. Scientists worldwide use the International System (SI) of measurement to measure work.

- The SI unit for force is newton (N).
- The SI unit for distance is meter (m).
- The SI unit for work is joule (J).

The equation to solve work problems looks like this:

Work = **F**orce x **d**istance or ***W = F* x *d***

Directions: Read the scenario and then complete the data table.

Scenario—A team of students investigated the relationship between the force required to move an object up an incline, the distance it moves, and the work done. Using a spring scale, the students lifted a box of marbles from the floor to the top of a table. They recorded the force needed to lift the box in the data table below. They repeated the experiment using a ramp and recorded the force needed to pull the box to the table top.

Distance Moved (m)	Force Required (N)	Work Done (J)
1 m (straight lift)	15 N	
2 m (ramp)	5 N	

Analyze the data and write a conclusion for the investigation.

____

____

____

____

____

____

Handout

Magnification Power

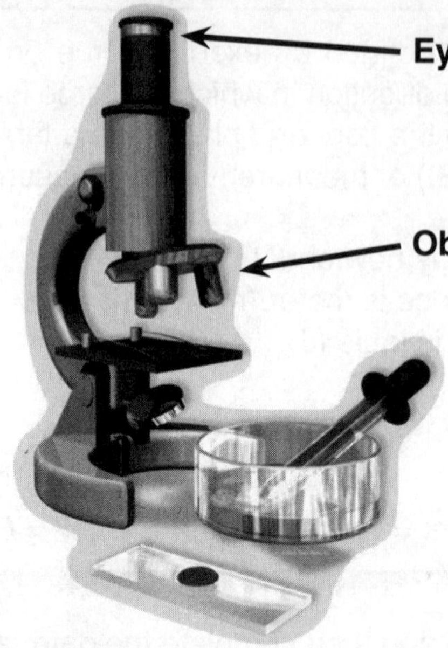

Lenses

Eyepiece/Ocular Lens: Your microscope will either have one or two ocular lenses. These are the lenses you will look through when examining a specimen with the microscope. The ocular lens is labeled on its side with 10X. This means that the ocular lens magnifies the image by a factor of 10 or 10 times.

Objective Lenses: These lenses allow you to change the magnification. Some microscopes have three objective lenses while others have four. The magnification is labeled on the side of each objective lens.

Magnification

Scanning objective lens: Not all microscopes have this lens. The scanning objective lens magnifies the image by a factor of 4. It is used to scan the slide to locate the specimen before viewing it at a higher magnification.

Low-power objective lens: This lens magnifies the image by a factor of 10.

High-power objective lens: This lens magnifies the image by a factor of 40.

Oil-immersion objective lens: This lens magnifies the image by a factor of 100. It requires a drop of immersion oil on the slide to provide good resolution.

Calculating Magnification

Formula: ocular lens power x objective lens power = total magnification power

The magnification power of a lens is usually indicated by a whole number followed by an "X". (Example: 10X) To calculate the total magnification used to view a specimen, multiply the ocular lens power times (x) the power of the objective lens (Example: 10X x 40X = 400X total magnification power).

Teacher Page

Unit: History of American Space Exploration

Goal: Students will be able to identify important people and events of American space exploration.

Common Core State Standards (CCSS):

ELA-Literacy.RST.6-8.2: Determine the central ideas or conclusions of a text; provide an accurate summary of the text distinct from prior knowledge or opinions.
ELA-Literacy.RST.6-8.7: Integrate quantitative or technical information expressed in words in a text with a version of that information expressed visually (e.g., in a flowchart, diagram, model, graph, or table).
ELA-Literacy.RST.6-8.9: Compare and contrast the information gained from experiments, simulations, video, or multimedia sources with that gained from reading a text on the same topic.

Materials List/Setup

Station 1: American Space Program—Summarizing a Speech (Activity); Speech—President John F. Kennedy, May 25, 1961 (Handout)
Station 2: Time Line—History of American Space Exploration (Activity); History of American Space Exploration (Handout)
Station 3: *Internet Hunt—Astronauts (Activity)
Station 4: *History of American Space Exploration in Video and Text** (Activity); History of American Space Exploration (Handout)

Activity: one copy per student
Handout: one copy per each student in a group

*Integration of Technology Skills

**Teacher Note: Below are the links to two possible online videos that could be used with the Station 4 activity. Remember to preview videos before showing to students.

50 Years of Exploration: The Golden Anniversary of NASA
<http://www.nasa.gov/multimedia/videogallery/index.html?media_id=11463021>
Kennedy Space Center's 50th Anniversary
<http://www.nasa.gov/multimedia/videogallery/index.html?media_id=147467071>

Opening Activity and Discussion Questions (Teacher-Directed)

1. If you had an opportunity, would you like to travel in outer space?
2. What do you know about the history of American space exploration?

Student Instructions for Learning Stations

At the learning stations, you will explore the history of American space exploration. Discuss your answers with other team members after completing each activity.

Closure: Reflection

Students will use the completed learning station activities to help compose the "Reflection: The Future of the American Space Program" writing activity.

Name: ______________________________ Date: ____________________

Station One: American Space Program—Summarizing a Speech

On April 12, 1961, Yuri Gagarin, a Soviet Union cosmonaut, became the first human in space. On May 25, 1961, President John F. Kennedy made a speech before a special joint session of Congress.

Directions: Read the handout, "Speech: President John F. Kennedy, May 25, 1961." Summarize the four national goals President Kennedy asked Congress to fund.

Goals	Summarize in your own words
One	
Two	
Three	
Four	

Name: ______________________ Date: ______________

Station Two: Time Line—History of American Space Exploration

Directions: Select ten events from the "History of American Space Exploration" handout. Add the dates and events to the time line. The first and last events have been done for you.

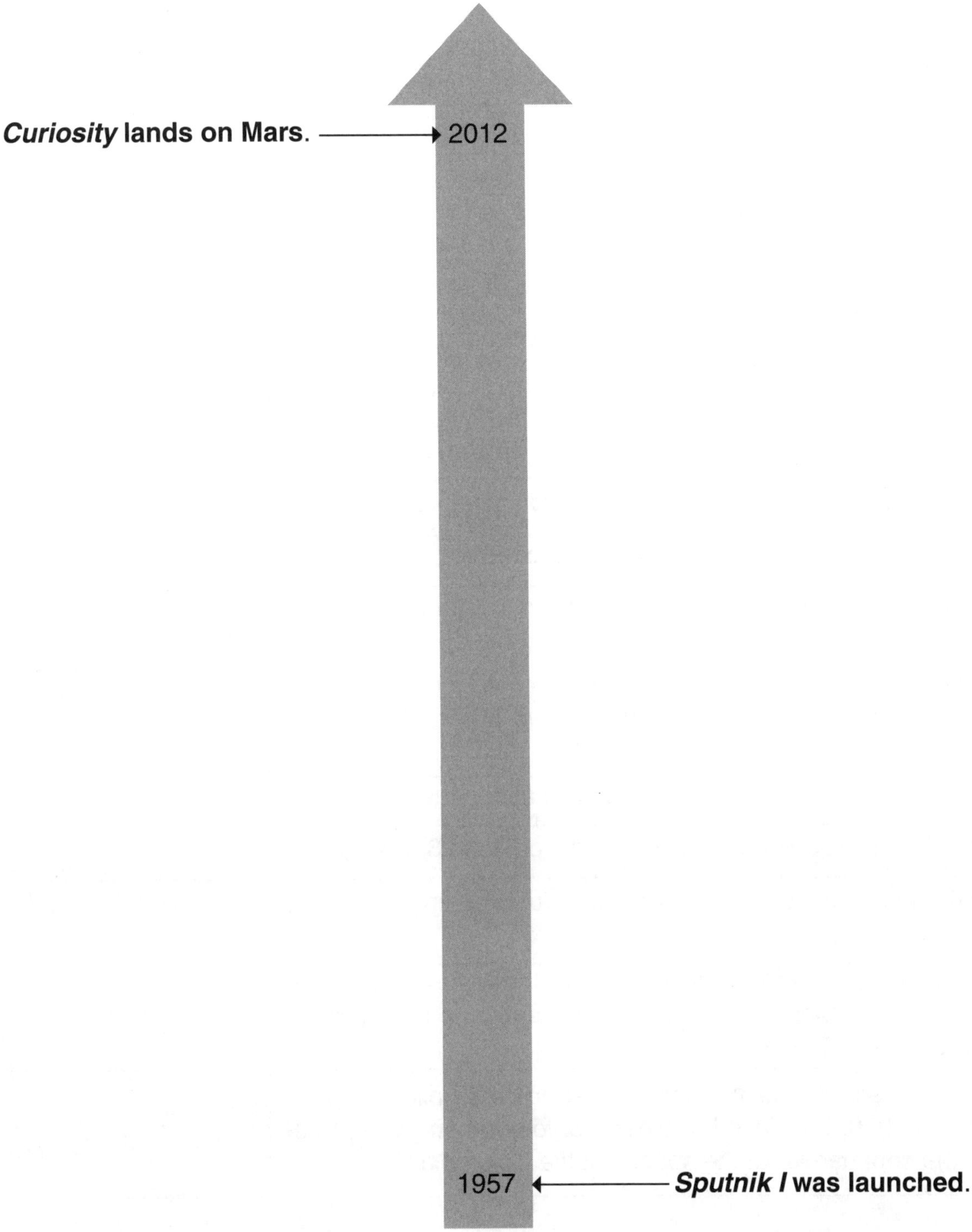

Name: ______________________ Date: ______________

Station Three: Internet Hunt—Astronauts

Directions: Use your Internet searching skills to help find the answers to the "Who Am I?" statements.

"Who Am I?" Statement	Astronaut
1. In 1959, I was chosen as one of the seven original astronauts in the *Mercury* program. I was the first American in space.	
2. On July 20, 1969, I became the first man to walk on the moon.	
3. I was born on November 22, 1942, in Philadelphia, Pennsylvania. In 1983, I was a mission specialist aboard the *Challenger* space shuttle. I was the first African-American to travel in space.	
4. In 1983, I became the first American woman in space.	
5. I was the commander of *Apollo 13*. A movie has been made of our nearly disastrous mission.	
6. I was a member of the *Apollo 11* crew in 1969. Along with Neil Armstrong, I was one of the first people to walk on the moon.	
7. I was a part of the *Apollo 11* mission, but I did not walk on the moon.	
8. In 1962, I became the first American to orbit Earth.	
9. I was the first African-American female to be admitted to the NASA astronaut training program.	
10. In 1959, I was one of seven astronauts selected for Project *Mercury*. On January 27, 1967, along with two other astronauts, I died during a pre-flight test at the NASA Space Center in Florida.	
11. In 1991, I became the first Hispanic female astronaut in the world.	
12. I performed the first American spacewalk on June 3, 1965.	
13. In 2007, I became the first teacher-astronaut to complete a space mission.	
14. In July of 2011, I was the commander of the last space shuttle mission. Our mission was to deliver supplies to the International Space Station.	
15. In 1995, I became the first female astronaut to pilot a space shuttle. In 1999, I achieved another first for women, when I became the commander of the space shuttle Columbia.	

Name: ______________________ Date: ______________

Station Four: History of American Space Exploration in Video and Text

Directions: Use the graphic organizer to compare the information gained from reading the "History of American Space Exploration" handout to the information gained from viewing the video on space exploration.

Video	Text

Name: ______________________________ Date: ______________________

Station Five: The Future of the American Space Program

On May 25, 1961, President John F. Kennedy addressed a special joint session of Congress on the future of American space exploration. In his speech, he stated, "I believe we should go to the moon." His goal was reached on July 20, 1969, when the American astronaut Neil Armstrong became the first human to walk on the moon.

Directions: What do you believe should be the primary focus in the future for the American space program? Give details or examples to support your answer.

Buzz Aldrin salutes the U.S. flag on the moon. This photo was taken by Neil Armstrong, July 20, 1969.

Handout

Speech—President John F. Kennedy, May 25, 1961

The following is an excerpt from a speech delivered by President John F. Kennedy on May 25, 1961, to a joint session of Congress.

I therefore ask the Congress, above and beyond the increases I have earlier requested for space activities, to provide the funds which are needed to meet the following national goals:

First, I believe that this nation should commit itself to achieving the goal, before this decade is out, of landing a man on the moon and returning him safely to the earth. No single space project in this period will be more impressive to mankind, or more important for the long-range exploration of space; and none will be so difficult or expensive to accomplish. We propose to accelerate the development of the appropriate lunar space craft. We propose to develop alternate liquid and solid fuel boosters, much larger than any now being developed, until certain which is superior. We propose additional funds for other engine development and for unmanned explorations—explorations which are particularly important for one purpose which this nation will never overlook: the survival of the man who first makes this daring flight. But in a very real sense, it will not be one man going to the moon—if we make this judgment affirmatively, it will be an entire nation. For all of us must work to put him there.

Secondly, an additional 23 million dollars, together with 7 million dollars already available, will accelerate development of the Rover nuclear rocket. This gives promise of some day providing a means for even more exciting and ambitious exploration of space, perhaps beyond the moon, perhaps to the very end of the solar system itself.

Third, an additional 50 million dollars will make the most of our present leadership, by accelerating the use of space satellites for world-wide communications.

Fourth, an additional 75 million dollars—of which 53 million dollars is for the Weather Bureau—will help give us at the earliest possible time a satellite system for world-wide weather observation.

Let it be clear—and this is a judgment which the Members of the Congress must finally make—let it be clear that I am asking the Congress and the country to accept a firm commitment to a new course of action, a course which will last for many years and carry very heavy costs: 531 million dollars in fiscal '62—an estimated seven to nine billion dollars additional over the next five years. If we are to go only halfway, or reduce our sights in the face of difficulty, in my judgment it would be better not to go at all.

Now this is a choice which this country must make, and I am confident that under the leadership of the Space Committees of the Congress, and the Appropriating Committees, that you will consider the matter carefully.

Handout

History of American Space Exploration

On October 4, 1957, the Soviet Union launched **Sputnik I**, the world's first artificial satellite. This event started the space race between the United States and the Soviet Union. In 1958, President Eisenhower established the National Aeronautics and Space Administration (NASA).

NASA became the agency responsible for the United States space program. Early in its history, a goal of NASA was to put an American in space. A year later Project *Mercury*, America's first human space flight program, was initiated. There were seven astronauts recruited for the project. On May 5, 1961, one of these astronauts, Alan Shepard, became the first American in space.

President Kennedy (right) visited Cape Canaveral in Florida in November 1963 to get an explanation of the *Saturn V* launch system.

On May 25, 1961, President John F. Kennedy announced his goal of landing a man on the moon and safely returning him to the earth. Kennedy's goal was met on July 20, 1969, when Neil A. Armstrong made his now famous remarks, "That's one small step for man, one giant leap for mankind" as he stepped onto the lunar surface during the *Apollo 11* mission. This mission was one of NASA's greatest achievements.

While humans have ventured no farther than the moon, they have used observational tools to travel to other planets. In 1974 and 1975, the United States' *Mariner 10* spacecraft repeatedly orbited the planet Mercury, photographing the surface of the nearest planet to the sun. *Voyager 1* and *2* were launched in 1977 and flew past Jupiter, Saturn, and Uranus and continued on toward the outer boundaries of the solar system and beyond. In April 1981, the space shuttle, the first true aerospace manned winged vehicle, was launched. The shuttle program was discontinued in 2011. In 1993, the *Magellan* spacecraft with the use of radar was able to view the Venusian landscape, revealing it as a desolate planet with a liquid atmosphere of carbon dioxide surrounding it. Since 1997, NASA has landed several rovers on Mars, which have discovered water and methane—both possible indicators of life on that planet.

Ongoing research continues to reveal new discoveries about our solar system and beyond. The Hubble Deep Field Telescope, launched in 1990, continues to gather information from the deepest reaches of space. The International Space Station, established in 1998, opened up a whole new universe of research possibilities and joint missions with other countries. In 2012, NASA successfully landed *Curiosity*, a land rover, on Mars. The rover is a robotic space probe used by NASA to collect data about the red planet.

What does the future hold for space exploration? Will it be to establish a colony on the moon or Mars? Will private citizens be able to book space flights? In the 1960s, Kennedy's goal of a moon landing seemed impossible. We now know that with science, developing technologies, and human ingenuity anything is possible.

Answer Keys

*If applicable, answers are provided.

Unit: Periodic Table of Elements
Summarizing Text (p. 4)

Main idea: The Periodic Table organizes the elements by their properties.

Detail: The Periodic Table is made up of horizontal rows called periods.

Detail: The vertical columns of boxes are called families.

Detail: Element key tells the element's name, symbol, atomic number, family, and gives other scientific information about the element.

Summary: The Periodic Table organizes the elements by their properties. The Table is made up of rows and columns. It is made up of horizontal rows called periods. The vertical columns of boxes are called families. The element key tells the element's name, symbol, atomic number, family, and gives other scientific information about the element.

Element Symbols (p. 5)

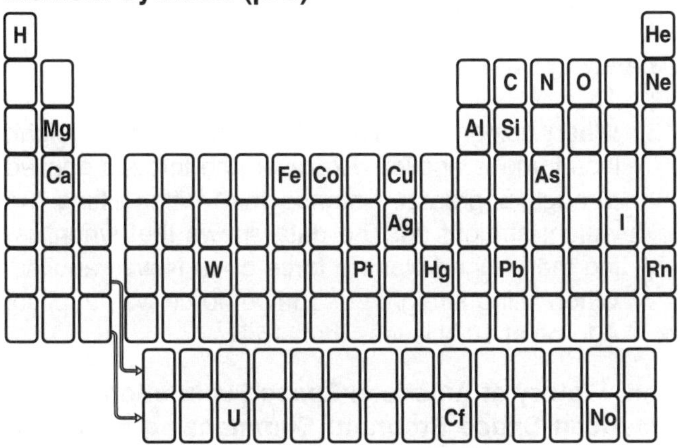

Locating Information From a Periodic Table (p. 6)

Oxygen
- Atomic Number: 8
- Symbol: O
- Family Number: 16
- Family Name: BCNO Groups
- Period: 2
- Solid, Liquid, or Gas: Gas

Bromine
- Atomic Number: 35
- Symbol: Br
- Family Number: 17
- Family Name: Halogen
- Period: 4
- Solid, Liquid, or Gas: Liquid

Gold
- Atomic Number: 79
- Symbol: Au
- Family Number: 11
- Family Name: Transition Metal
- Period: 6
- Solid, Liquid, or Gas: Solid

Graphing Textual Information (p. 7)

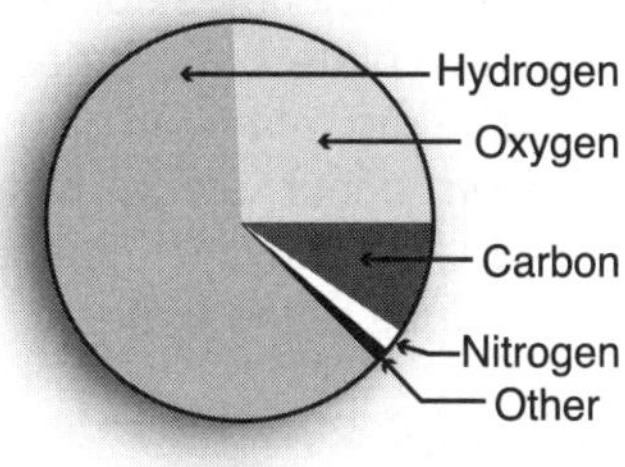

Unit: Electricity
Text Features (p. 12)

1. *Answer*—In conduction, charges are transferred by direct contact with an object. With induction, charges can be induced without direct contact.
   Text Feature—bullets
2. *Answer*—positive charge
   Text Feature—caption
3. *Answer*—dry
   Text Feature—heading
4. *Answer*—Lightning is a discharge of static electricity.
   Text Feature—sidebar

Following a Multi-Step Procedure (p. 13)

Materials: one empty soda can; one air-filled balloon

Procedure: 1. Place an empty aluminum soda can on its side on a smooth surface.
2. Rub the balloon on your hair.
3. Place the balloon near the side of the can at a low angle.
4. Without touching the can, use the balloon to push the can along.

Results: (Answers may vary.) The balloon should make the can move.

Creating a Circuit (p. 16)

Part A—Results: Answers may vary, but students should report a slight tingling to the tongue.

Part B
1. A series circuit was created.
2. The lemon is the battery. The switch is your tongue.
3. You would need to add more lemons.

Constructing a Model (p. 17)

1. It is a series circuit because the electrical current flowed through all the devices in the circuit in just one path to make a complete circuit.
2. The paper clip acted as the switch. The switch is used to open or close the circuit.

Unit: Volcanoes

Plate Tectonics: Analyzing Textual and Visual Information (p. 22)

1. The motion of magma in the mantle (asthenosphere) just under the crust causes the movement of the plates. The plates sometimes collide and grind past each other at the plate boundaries. Earthquakes, volcanic activity, mountain-building, and oceanic trench formation occur along plate boundaries.
2. Answers will vary, but may include the names, location, and boundaries of the plates

Volcano Vocabulary (p. 23)

1. cool, slow-moving lava that forms spiky chunks
2. broad crater three times wider than it is deep
3. type of volcano with narrow bases and steep sides
4. magma that has reached the surface
5. hot, fast-moving dark lava with iron and magnesium

Volcanoes in the United States (p. 24)

(Date of Latest Activity may change)

1. Stratovolcano; 2005
2. Shield with Caldera, Cones; 1937
3. Stratovolcano with Caldera; 1992
4. Cinder Cone; 1250 A.D.
5. Stratovolcano; 1786
6. Volcanic Field; Present hydrothermal
7. Caldera; 1750s
8. Shield; 1984
9. Shield; 1750
10. Cinder Cone; About 2,000 years ago
11. Stratovolcano; 1866
12. Caldera; About 2,000 B.C.
13. Stratovolcano; 1880
14. Stratovolcano; 1825
15. Stratovolcano; 2005

Mt. Saint Helens (p. 25)

Answers will vary.

1. flattened 230 square miles, ash covered everything downwind, mud and debris flows took out bridges and homes
2. GPS, infrared imaging, acoustic flow sensor, Doppler radar
3. Hazards from lava flows to explosive eruptions, mud flows, debris flows, debris avalanches, and airborne volcanic ash
4. Hawaii, Alaska, The Cascades, Yellowstone, and Long Valley in California
5. Answers will vary.

Unit: Science Lab

Lab Safety (p. 30)

1. safety goggles: proper eye protection should be worn
2. lab apron: wear to protect clothing and body
3. poison: substance may be poisonous if touched, inhaled, or swallowed
4. electrical: possible electrical shock or burn
5. flame: open flames, may cause fire
6. heat-resistant gloves: hand protection needed when handling hot or extreme cold
7. sharp objects: tools or glassware that can cut
8. fumes: harmful vapors, work in ventilated area

Lab Equipment (p. 31)

1. graduated cylinder
2. ring stand
3. horseshoe magnet
4. spring scale
5. microscope
6. safety goggles
7. pipette
8. test tube
9. forceps
10. triple-beam balance
11. magnifying glass
12. beaker

Using a Compound Microscope (p. 32)

Answers will vary for questions 1–6.

7. Slide appears to move left; slide appears to move to the right
8. Slide appears to move down or toward you.
9. It appears upside down.
10. In a microscope, the image is actually inverted sideways and upside down, like a double reflection.

Analyzing Data (p. 35)

1. 15 J
2. 10 J
3. When using an inclined plane to move a weight, less force is applied. However, the force is applied through a greater distance than when lifting the weight straight up. The data shows that when using the inclined plane a force of 10 N was applied. When lifting straight up, 15 N of force was applied. A force of 10 N is less than 15 N

Unit: History of American Space Exploration

American Space Program: Summarizing a Speech (p. 38)

1. develop the technology needed to land a man on the moon and return him safely to the earth.
2. accelerate development of the Rover nuclear rocket
3. accelerating the use of space satellites for world-wide communication
4. a satellite system for world-wide weather observation

Internet Hunt: Astronauts (p. 40)

1. Alan Shephard
2. Neil Armstrong
3. Guion S. Bluford
4. Sally Ride
5. James A. Lovell, Jr.
6. Buzz Aldrin
7. Michael Collins
8. John Glenn
9. Mae C. Jemison
10. Gus Grissom
11. Ellen Ochoa
12. Ed White
13. Barbara Morgan
14. Christopher Ferguson
15. Eileen Collins

History of American Space Exploration in Video and Text (p. 41)

Answers will vary depending on the video the teacher decides to use with this activity.